D1164566

Even arid, or dry, earth supports life

ELEMENTS

Earth

Aaron Frisch

A⁺
Smart Apple Media

COPYRIGHT

Published by Smart Apple Media

1980 Lookout Drive, North Mankato, MN 56003

Designed by Rita Marshall

Copyright © 2002 Smart Apple Media. International copyright reserved in all countries. No part of this book may be reproduced in any form without written permission from the publisher.

Printed in the United States of America

Photographs by KAC Productions (Kathy Adams Clark, Larry Ditto, Bill Draker), Tom Stack & Associates (Milton Rand, Greg Vaughn)

Library of Congress Cataloging-in-Publication Data

Frisch, Aaron. Earth / by Aaron Frisch. p. cm. – (Elements series)

Includes index.

ISBN 1-58340-074-5

1. Earth–Juvenile literature. [1. Earth.] I. Title. II. Elements series (North Mankato, Minn.)

QB631.4 .F75 2001   550–dc21   00-067973

First Edition 9 8 7 6 5 4 3 2 1

Earth

CONTENTS

Inside the Earth

The earth was formed about four and a half billion years ago. Since that time, wind and rain have pounded at the earth, shaping and reshaping it. Rocks and mountains have risen from the earth and crumbled, forming new rocks and soil. Ever so slowly, these changes molded the surface of the earth as we know it. Scientists who study the earth are called geologists. They learn about the earth by studying rocks from different time periods. They also study underground

Layered rock formations

vibrations during **earthquakes** to find out more about the inside of the earth. Geologists think that there are three layers of rock that form the earth. The outside layer is called the crust. The surface of the earth is part of the crust. The crust is fairly thin—only about five miles (8 km) thick in some places. Under the crust is the second layer, called the mantle. The mantle is a thick layer of hot rock. Under the mantle is the third layer, called the core. The outer core is made up of melted rock

Gravity is the force that pulls objects toward the center of the earth. Gravity helps to even out the earth's surface by pulling soil and rocks to low areas.

called magma. The very center of the earth, called the inner

core, is a huge ball of a solid rock called iron. Geologists think

the inner core may be as hot as 13,000° F (7,200° C)!

Magma erupting through the earth's crust

The earth's crust is covered by rocks, plants, and water

Looking at Rocks

The formation of rocks is a very slow process that takes millions of years. All rocks are made of **minerals**. Minerals are in turn made up of basic substances called elements. Near the surface of the earth are lighter elements such as silicon and oxygen. Rocks that contain

People use rocks for many things. Some hard kinds of rock are used to build houses. Other types of rock are used to make cement.

heavier elements, such as iron and aluminum, are found deeper in the earth. Sedimentary rocks formed when sand, dirt, and dead plants and animals were carried into valleys by

wind and rain. Slowly, this material piled up, layer by layer, and

hardened into rock. These rocks tell geologists a lot about the

earth's history. Many sedimentary rocks contain **fossils**. By

A prehistoric leaf fossil

studying fossils, geologists can figure out how old rocks are

and when different plants and animals lived.

What Is Soil?

Much of the earth is covered by soil. Soil is a mix-

ture of tiny bits of rock and dead matter from plants and ani-

mals. It is formed as natural forces such as rain and wind grad-

ually wear away at rocks. This process is called erosion.

Glaciers help this process by picking up rocks and grinding

them into pebbles and sand as they slide across the land. As

water soaks into these crushed bits of rock, plants begin to

grow from them. Soil is very important to life on Earth.

Trees and plants anchor their roots in soil and draw water and

nutrients from it. When plants and large animals die, worms

Rocks in different stages of erosion

and tiny animals called microbes in the soil break the dead matter down into nutrients. New plants use these nutrients to grow, and the cycle continues. Soil that contains bigger bits of rock looks and feels like sand. Water drains quickly through sandy soil. Soil that contains much smaller bits of rock is called clay soil. Clay soil

The best soil for growing plants is called topsoil. Topsoil is usually a dark color and makes up the top few inches of soil.

is tightly packed and can hold water for a long time. Soil can be black, brown, red, or yellowish in color. In some places, it

Wet soil is ideal for rice farming

may be only two inches (5 cm) deep. In other places, the

earth's surface may be covered by 10 feet (3 m) of soil.

Protecting the Earth

Soil is very valuable to people, since we need it to grow

the food we eat. But soil easily washes away from the land if it

is not properly cared for. To protect it, ranchers and farmers do

not let cows and other animals graze too long in one area.

Grass and plants protect soil from rain and wind. If animals eat

all the plants, the soil may be blown or washed away. Farmers

A rancher's cattle grazing in a grassy pasture

also add **fertilizers** to fields. This adds more nutrients to the soil, helping farmers grow better crops. The surface of the earth is always changing. Rain falls on the earth and runs into rivers and streams. The rivers and streams then flow across the earth, wearing away at the rock. We cannot stop these and other natural changes from happening, but we must take special care to protect the earth's soil so that it will always be able to support life.

Prairie dogs burrow in the soil

Soil and Water

There are many different types of soil. You can see some differences for yourself with a simple experiment.

What You Need

A pencil

Two paper cups

Some sandy soil

Some clay soil

Two measuring cups

Water

What You Do

1. Use the pencil to punch a hole in the bottom of each paper cup. Fill one cup with sandy soil and the other with clay soil.

2. Hold the cup with sandy soil over an empty measuring cup. Pour one-half cup (120 ml) of water into the soil. Measure how much water drains out in two minutes. Pour out the water.

3. Hold the cup with clay soil over an empty measuring cup. Pour one-half cup (120 ml) of water into the soil. Measure how much water drains out in two minutes. Compare this measurement to the first one.

Clay soil is packed tightly together, so water cannot get through easily. Sandy soil is packed more loosely, so water drains through much more quickly.

The sandy soil of a desert

INFORMATION

Index

Words to Know

earthquakes (ERTH-kwakes)—shaking of the ground caused by shifting rock inside the earth

fertilizers (FER-tuh-ly-zers)—substances such as animal droppings or man-made chemicals that help soil support plants

fossils (FAH-suls)—preserved remains of prehistoric plants and animals

glaciers (GLAY-shers)—huge sheets of ice that move very slowly over land

minerals (MIN-uh-ruls)—basic particles found in rocks and other non-living substances

nutrients (NOO-tree-ents)—tiny substances that help living things grow

Read More

Brimner, Larry Dane. *Earth.* New York: Children's Press, 1998.

Gibbons, Gail. *Planet Earth: Inside Out.* New York: Morrow Junior Books, 1995.

Green, Jen. *Earth.* Brookfield, Conn.: Copper Beech Books, 1998.

Internet Sites

EarthNet

http://agc.bio.ns.ca/EarthNet/

U.S. Geological Survey Home Page

http://www.usgs.gov/

Elements Online Environmental
Magazine
http://www.elements.nb.ca/index1.htm